I0213459

In the Shadow of the Battleship:

Considering the Cruisers of World War II

RICHARD WORTH

NIMBLE BOOKS LLC

NIMBLE BOOKS LLC

ISBN-13: 978-1-60888-076-8

Copyright 2008 Richard Worth
Nimble Books LLC
1521 Martha Avenue
Ann Arbor, MI 48103-5333
http://www.nimblebooks.com

All photos, except as credited, are from the author's collection.

The body text and headings in the book are in the Constantia font, designed by John Hudson for Microsoft.

Front Cover: *Georges Leygues* leads a column of French light cruisers. (www.history-on-cdrom.com)

Contents

NIMBLE BOOKS LLC

ACKNOWLEDGMENTS

The author expresses his thanks to William J. Jurens, John Jordan, Sander Kingsepp, Arseny Malov, Stephen McLaughlin, and Enrico Cernuschi for important technical assistance.

Also, the Boris Lemachko Collection (lemachko@mtu-net.ru) and History on CD-ROM (www.history-on-cdrom.com) have kindly provided photographs for this volume.

ABOUT THE AUTHOR

Richard Worth, author of *Fleets of World War II* (Da Capo: 2001) and *Raising the Red Banner: The Pictorial History of Stalin's Fleet 1920-1945* (Spellmount: 2008), is also a regular contributor to the *Warship International* and *Warship* journals.

FOREWORD

In a world of 35,000-ton battlewagons hurling 1-ton shells over the ocean waves, a set of 10,000-ton jacks-of-all-trades firing 280-lb shells don't evoke the same swells of testosterone and poetry—hardly a surprise, despite the fact that cruisers performed vital services in all theaters of World War II, maintaining a pace too frenetic for any battleship to rival, enduring losses too steep for any battleship fleet to sustain.

With their lesser marketability, cruisers have earned less ink, meaning fewer worthwhile references and more enduring misconceptions. Design details remain elusive, and the information vacuum naturally fills with an over-reliance on statistics which, in their most commonly encountered uses, provide heaps of numbers but little actual information.

The problems quickly compound because cruisers, as multi-role platforms, present an inherently more complex subject than battleships. Cruisers crowd within a general set of relatively powerful ships capable of independent operations; but within that broad grouping lie many incarnations, with designs tailored to fit specific missions, environments, and finances by navies dealing with specific political forces and vocabularies.

Vocabulary is a prime cause of cruiser confusion. For example, the phrase "treaty cruiser" implies a connection to the Washington Treaty's definition of a cruiser; in fact the treaty never mentions cruisers, some designs described as treaty cruisers actually precede the treaty, and each navy had its own opinion of treaty restrictions. Even the word "cruiser" has its ambiguity; at the margins, classification begins to blur, and some "cruisers" look more like destroyers or battleships. In reviewing the ships of World War II, the essays in this volume focus on the ships routinely labeled as light cruisers and heavy cruisers and, it is hoped, dispel some of the confusion.

Liars and Statistics: Protection in Cruiser Design

Few things cause more confusion about warships than facts. Reference books overflow with facts, and the most pernicious appear in the form of statistics. The numbers make it plain that Italy's cruiser *Bartolomeo Colleoni* (capable of 37 knots) could easily outrun Australia's *Sydney* (capable of 32.5 knots), yet the Battle of Cape Spada irks us with a different reality.

Some statistics, like so many found in an old *Jane's*, are simply wrong; but even the accurate ones can mislead, appearing in the vacuum of publication with no explanation of the qualifying factors. A survey of statistics for cruiser armor serves as a perfect case study. When a book dutifully declares that HMS *Such-and-Such* had an *x*-millimeter belt, it actually reveals very little unless it also describes the placement of the armor, the extent of its coverage, and the qualities of the plating. A country-by-country overview shows how these factors come into play.

United States

Many folks have no doubt that the best armor protection belonged to America's cruisers. The brawny *Baltimore* class went to sea with turret armor 8 inches thick, an inch thicker than any armor aboard any foreign cruiser. The conning tower design featured 6-inch sides, matching the belt armor and equaling the thickness in the best foreign competitors. But reality steps in to parse these figures. The 8-inch turret armor has no peculiarly American qualifier; it covers only the front of the gunhouse while thinner plates covered the other surfaces, as usual among the world's cruisers. The conning tower, however, involves a quirk. On the minus side, the first six members of the class never received the armor; on the plus side, subsequent units not only had the armor, but at a thickened 6.5-inch maximum.

Belt protection is where American statistics look deceptively good. *Baltimore*'s 6-inch armor tapered at its lower edge to only 4 inches—but many foreign cruisers had a lower-edge taper, too. More importantly, *Baltimore*'s shallow belt covered only the midships to protect the machinery spaces. The area forward of the bridge had no waterline belt whatsoever, only some thinner plating below the waterline to guard against submerged hits.

The situation aft is even worse. Not only is the armor entirely below the waterline, but it is recessed into the ship's interior, not so much a belt but an armored bulkhead, and only 3 inches thick. With the side armor submerged, the deck armor must also be submerged—the hull above is utterly exposed. This then is the key principle regarding American cruisers: the admirable thicknesses come at the cost of leaving extensive hull volume unprotected. Though the details varied, all of America's treaty-era designs shared this principle, even those whose armor was much thinner than *Baltimore*'s. Only with the massive postwar designs—the *Des Moines* and *Worcester* classes— does thick plating combine with widespread coverage.

The complications continue. Some ships had a slight tumblehome, angling the belt for easier penetration. Statistics given for American cruiser belts often include backing plate thickness as though it were part of the belt. While the U.S. made its backing plates from armor-grade metal (often 0.75-inch), laminations are inferior to single plates; a 2.25-inch belt on a 0.75-inch backing plate doesn't equal a single 3-inch plate. American decks were often laminated as well.

Elderly *Omaha* warrants special mention. Credited with a 3-inch belt and a 1.5-inch deck, she seems reasonably protected. In fact, this plating enclosed two small boxes, one for the machinery and one for the steering gear, and all the armor was laminated. Apart from the wiring tube extending to the conning tower (1.25-inch), there was no other plating thicker than half an inch. The magazines had no armor. The enclosed gun mounts provided shelter from the elements but, with only 6.4mm plating, would not keep out small shells or even splinters.

Later cruisers had family traits of a more desirable sort. The steering gear enjoyed protection similar to that given the machinery, sometimes even better. And the quality of American armor for defeating cruiser-caliber gunfire was unsurpassed. During the construction of the *Brooklyn*-class light cruisers, the Americans began a switch over to face-hardened armor for the thicker vertical plates. Some navies used only homogenous armor for their cruisers, and normally the difference between face-hardened armor and homogenous armor might not matter for cruisers. However, the Japanese (who never used face-hardened cruiser armor) specialized their cruiser shells for performance against homogenous plates. Consequently, at the Battle of Savo Island, Japanese shells went straight through the 8-inch homogenous armor of *New Orleans*-class turrets; but at the Battle of Cape Esperance, *Brooklyn*-class *Boise* benefited from her face-hardened plating. An 8-inch shell hit her 5-inch belt with enough force to drive a 2-inch dent into it, but otherwise caused only limited damage. Another shell struck a 6.5-inch faceplate and failed so badly that the Americans concluded it had to be a non-AP shell with a windscreen over a flat nose; in fact, that description matches Japan's specialized AP shell. A third shell struck the 6-inch barbette armor and managed to burrow into it, jamming the turret; but in the process, the shell snapped apart so completely that it couldn't burst.

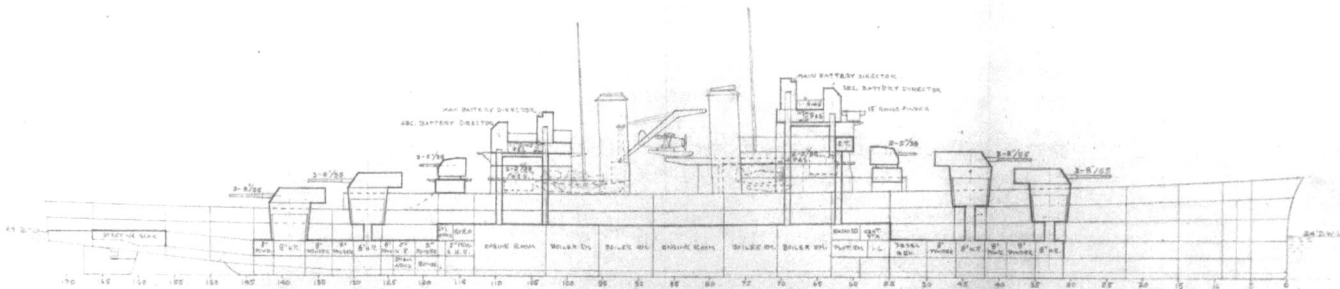

Figure 1. Sketch for an American heavy cruiser design. Even in such a massive ship—17,000 tons—the hull had little armor above the waterline.

JAPAN

Designed during the Washington Treaty negotiations, the *Furutaka/Aoba* class resembled later American types with their midships-only belts, but significant differences became trends toward a unique Japanese system. The belt armor had a 9° inclination to improve its shell resistance, a trait increasingly pronounced in subsequent heavy cruisers. Ultimately the *Tone*s had their belts inclined at 20°, and this armor extended in a long taper all the way to the ship's bottom, though this meant that the lower portion of the armor was internal—some hull volume lay outside its protection. Deck armor over the magazines was mounted low by the waterline as in American ships, but it generally covered all the way to the ship's side. The deck over the machinery had an usual feature, a slanted area along each side, angling down to meet the top of the belt. This "turtledeck" can cause confusion. A statistic stating simply that *Tone* had 65mm deck protection fails to convey that this covered only the slope, while the midships horizontal armor was just 31mm. The slope needed its double thickness because it was angled to face incoming shells more directly (and because of the mediocre quality of Japanese cruiser armor).

Figure 2. The gun shields on Japanese light cruisers like *Tenryu* gave little protection in combat. (courtesy of the Boris Lemachko Collection)

Another feature resembling American practice was the provision of significant thickness guarding the rudder gear. However, while the

Americans reserved their thickest armor for the turrets, the Japanese never put more than 25mm on their gunhouses, sufficient as splinter protection but inadequate even against 5-inch shells. The barbettes typically had a similar thickness, though the coaming over the armor deck was stout to shield the ammunition hoist openings.

The old light cruisers, contemporary with *Omaha*, resembled her in skimpiness of armor. *Omaha* was actually superior in that the Japanese ships had no protection for their steering, yet their proportion of armored waterline was greater. They had thicker gun shields but, at 10mm, still inadequate. The later *Agano*s brought significant advances. The rudder had some protection, and the 19mm gunhouse plating could counter most splinters from light cruiser shells. The magazine protection was set inboard from the outer plating. *Oyoda* had a similar arrangement, adjusted for her all-forward main battery.

FRANCE

Skimpy as Japanese turret armor may seem, the first French light cruisers had no protection anywhere thicker than 30mm; then to make matters worse, the first French heavy cruisers inherited this 30mm maximum—truly sad, given that this 30mm "armor" was actually two layers of high-tensile steel. The subsequent *Suffren*-class heavy cruisers present some tricky issues. The four ships introduced progressively heavier protection, certainly a good thing, but designers applied the armor in an uncoordinated patchwork, guarding only a limited volume and showing numerous gaps. Then came *Algérie*. Her 80mm deck was the thickest deck protection of any wartime cruiser. She also had a respectable 110mm belt. In some ways, the true level of protection exceeded these figures as the French made frequent use of thickened bulkheads behind the belt, also retaining compartments of coal purely for protective reasons—like the thickened bulkheads, coal could stop splinters, if not intact shells. On the negative side, the 80mm deck thinned to only 30mm in some sections—a strange and vulnerable feature.

Meanwhile, armor had continued as an afterthought in light cruiser designs until *La Galissonnière* entered service with a 105mm belt and a 38mm deck. With a layout simpler than *Algérie*'s, the percentage of protected volume slightly exceeded that of American ships. Ton for ton, *Algérie* and *La Galissonnière* rate among the best cruisers of the war.

Figure 3. *Dupliex* of the *Suffren* class, whose gappy armor gave limited protection. (courtesy of History on CD-ROM)

ITALY

Like the French, the Italians began with lesser emphasis on armor. Their *Trento*-class heavy cruisers had a simple armor scheme with a 70mm belt and a 50mm deck; but while sometimes derided as "tinclads," they in fact had an unusually large portion of protected volume. Designers proceeded from this start to the fantastic *Zaras*, which rate as the best-armored cruisers of the war. The 150mm main belt and 70mm armor deck enclosed a large volume, while additional plating in the upper hull made for greater complexity than in *Trento*'s design. Turrets, barbettes, and conning tower armor also reached a maximum of 150mm, face-hardened armor of high quality. The worst news in all this was the skimpy plating afforded the steering gear, an Italian family trait. Politics dictated that the next and final Italian heavy cruiser would resemble *Trento* rather than *Zara*, but at least this *Bolzano* retained the virtue of large protected volume.

Meanwhile the light cruisers had gone through their own evolution. The first series of what would be dubbed the "Condottieri" classes had protection intended only to localize damage from French *contre-torpilleur* guns; except for the conning tower and its 40mm maximum, the thickest plating was the 24mm belt. Behind this was an 18mm bulkhead adequate against 138mm shell splinters. But facing the reality of French cruisers with 152mm and 155mm guns, "Condottieri" standards improved. At first the basic layout of belt and bulkhead continued, but thicknesses increased, and *Duca d'Aosta* had the same 70mm belt as a *Trento*. Then things changed dramatically in the final class, the *Duca degli Abruzzi*s. Instead of mounting a bulkhead behind the belt, designers set a 30mm plate outboard of the main 100mm belt. The outer plate was intended to strip the armor-piercing cap off incoming shells; these naked shells then had to confront the main belt. In this way, Italian light cruisers progressed from some of the worst-protected designs to some of the best.

Figure 4. By the time of *Eugenio di Savoia*'s construction, Italy had progressed from the minimal armor of the early "Condottieri" classes. (courtesy of History on CD-ROM)

GREAT BRITAIN

Britain went into World War II with numerous World War I leftovers. The "C" classes featured belt "armor" as thick as 3 inches—actually a two-layer lamination of high-tensile steel rather than a single layer of genuine armor. These 3 inches protected only the midships, thinning progressively while extending toward the ends. The rudder had a 1-inch deck overhead, but the rest of the hull had only partial deck coverage with a stretch of 1-inch plating running along each flank. The "D" and "E" classes followed this model but with some improvements; the gun shields thickened to 1 inch, and the magazines had individual boxes of splinter protection. The large *Hawkins*-class ships were similar but with more extensive deck protection and the first instance of inclined belt protection for a cruiser.

After the Washington Treaty, there came the heavy cruisers with their routinely misunderstood armor. Designers repeated the magazine boxes but eliminated all belt protection. The result was extreme concentration; the

IN THE SHADOW OF THE BATTLESHIP

armor for the ammunition spaces was formidable, thick enough for face-hardening, but the rest of the ship was virtually naked. In other words, the armor did little to keep the ships afloat except to prevent magazine explosions. The main turrets repeated the 1-inch standard of protection, but in contrast to Japanese design, but the hoists had no barbette protection to speak of. The steering gear relied on 1.5-inch plating. Fortunately, most units did eventually receive belt armor amidships, but some went through their careers with just an inch of metal down their sides.

Light cruiser construction then resumed, featuring belt protection for the midships, boxes for the magazines, and little protection elsewhere. But the standards improved during the "Town" classes, and external belts at last covered the entire citadel, providing a respectable percentage of protected volume. Turret armor thickened, but barbettes, conning tower, and steering protection showed lower priority. Subsequent 6-inch cruisers followed this pattern through war's end, though the little *Dido*s reverted to a midships-only belt and minimal gunhouse protection.

GERMANY

German designers worked under burdensome conditions, not limited merely by the Washington and London Treaties, but by the Versailles Treaty. Their first new cruiser design, *Emden*, looked much like their old cruiser designs, including an old-fashioned turtledeck armor scheme. This had the deck sloping down with a 20mm thickness behind the 50mm belt in hopes of minimizing damage after belt penetrations, but leaving the hull volume above the slope unprotected. It also opened the possibility of a shell arching over the belt to hit the slope at a favorable angle to penetrate when a horizontal deck might have defeated the shell. The next classes, the "K" class and the "pocket battleships," both had horizontal decks while thickened bulkheads assumed the job of backing the belt protection.

Belts were thin in German cruisers, but the 80mm figure usually given for the pocket battleships oversimplifies an extremely complex arrangement, further complicated by the fact that the three sister ships differed significantly in their details. The 80mm thickness covered only the midships, and only partly—there were actually two strakes, one 80mm and the other

Figure 5. A close look reveals the limited extent of _Leander_'s belt protection.

only 50mm. Abreast the magazines, a different set of plates gave 60mm of protection (or 100mm, depending on the individual ship). In all cases, an inclination of about 13° increased the resistance of the plates but also left some hull volume outside the protection of the belt. The height of the belt, as usual for German ships, gave good coverage.

Strangely, subsequent German light cruisers reverted to turtledeck layouts, though benefiting from belt inclination. Even the massive *Hipper*-class heavy cruisers had a turtledeck as demanded by Admiral Raeder himself, perhaps in response to the modest belt armor—80mm inclined at 12.5°. What *Hipper* had in her favor was a family trait of a stout conning tower (a face-hardened 150mm—even the light cruisers typically had 100mm for their conning towers). The pocket battleships had carried thick turret protection, and *Hipper* improved on that (at a maximum of 160mm, second only to American designs) while shaving the armor deck to just 30mm on the flat portion.

Figure 6. *Graf Spee* had a smarter armor layout than the later and larger *Hipper*s.

Soviet Union

The Soviets went through the war with a hodgepodge: antique protected cruisers, light cruisers designed before World War I but completed after World War I, and a few modern ships driven through the construction process. The protected cruisers barely merit a thought, having no belt armor whatsoever but only a protective deck incapable of withstanding modern cruiser gunfire. While the light cruisers could not be mistaken for modern units, they had respectable standards of protection with a fully horizontal deck sitting atop the 75mm belt armor. The belt was unusual in having face-hardened plating, which rarely appeared in naval armor in thicknesses less than 100mm. The modern cruisers had a simple armor scheme, complicated only by the tapering of the belt armor at its top edge, a truly odd feature. The extreme difficulties with quotas and quality control which Soviet industry experienced at this time do not hint at high standards of armor quality.

Given the limitations inherent in statistics, a true understanding of cruiser armor requires a look at armor schematics, which portray the realities with precision that exceeds even the details in this essay. Strangely, schematics for Axis ships are much easier to come by than schematics for Allied ships, which continue to suffer inexplicable neglect in published works. The Appendix on Further Reading gives a list of useful resources with special emphasis on armor drawings. The careful reader will keep in mind that real-world ruggedness goes beyond armor, involving a lengthy list of additional factors: machinery arrangement, roller path placement, subdivision, emergency power, pumps, and so forth. All of this leads well away from the seeming objectivity of armor statistics, and properly so.

Figure 7. The Russian *Svetlana*-class light cruisers had a respectable armor scheme. This view shows the area where belt plating was removed in converting a ship for mercantile service. (courtesy of the Boris Lemachko Collection)

QUARTS INTO PINT POTS: THE BEST OF THE TREATY CRUISERS

When the Washington Treaty placed restrictions on ships displacing more than 10,000 tons or mounting guns larger than 8-inch, it ostensibly established the maximums for cruiser design. In reality, it lured all the signatories to view those criteria as minimums lest foreign rivals establish a qualitative edge. With the five navies building to a common criterion, historians would seemingly have a solid basis for comparison.

In fact, the treaty provided only one criterion, and with each navy emphasizing its unique needs, the concept of "best" acquires some elasticity, especially given the willingness of some nations to ignore the treaty limits. With the seas thus populated by a variety of designs, a close look is required to see how each class succeeded on its own terms.

THE MEDITERRANEAN

The natural rivalry between France and Italy expressed itself in the fact that each fleet built seven heavy cruisers. While both countries planned for a naval duel in the western Mediterranean, there were significant differences in their ideas on cruiser design. France had more extensive overseas commitments and the prospects of an Atlantic scenario against the Germans. The Italians had begun to look ahead to a time when fleet operations outside the Mediterranean became viable, but the requirements of a "Breakout Fleet" didn't figure in their treaty cruiser designs.

The French generally honored the treaty rules, with minimal fudging. Their grandest infringement lay in their use of coal; fuel lay outside the treaty limits, but as some French ships lacked coal-fired boilers, their coal could serve only as protection—an overweight of perhaps 600 tons. The Italians' willingness toward outright violation sent them reaching past 11,000 tons, breaking the letter and the spirit of the law and making some excellent ships.

DUQUESNE

> Displacement: 10,000 tons[1]
> Range: 5000nm at 15 knots
> Speed: 34 knots
> Armor: 0.9-inch belt, 0.9-inch deck, 1.2-inch turret
> Main Battery: eight 8-inch guns, 2363-lb broadside

SUFFREN

> Displacement: 10,000 tons
> Range: 4600 nm at 15 knots (oil); plus 2000nm at 11 knots (coal)
> Speed: 32 knots
> Armor: 2-inch belt, 1-inch deck, 1.2-inch turret
> Main Battery: eight 8-inch guns, 2363-lb broadside

ALGÉRIE

> Displacement: 10,000 tons
> Range: 8000nm at 15 knots
> Speed: 31 knots
> Armor: 4.3-inch belt, 3.1-inch deck, 3.9-inch turret
> Main Battery: eight 8-inch guns, 2363-lb broadside

Routinely ranked among the worst of treaty cruisers, France's two *Duquesne*s certainly had the lowest level of protection; the statistics actually flatter them. On the plus side, they had powerful guns, and their excellent sea-keeping made them superior gun platforms. They showed a high degree of watertight subdivision. However, their range hardly matched the requirements of Atlantic duty, nor did habitability standards suit them to lengthy missions. Their 3-inch secondary guns presented little threat to any enemy, surface or aerial.

The next class, the four *Suffren*s, incorporated growing emphasis on protection, as shown by the weight dedicated to protection in each unit:

---

[1] Those who have read "Liars and Statistics" will appreciate that statistics have limited value, but they can serve as general indicators, especially for ships built by the same navy.

Suffren 635 tons (50% more than in *Duquesne*), *Colbert* 750 tons, *Foch* 1352 tons, *Dupleix* 1528 tons. By adding more armor and concentrating it in the ship's interior, the later units had fewer gaps, giving a realistic hope to defeat at least some 6-inch shells before they reached the vitals. Many other *Duquesne* features recurred in *Suffren*, though her three sisterships switched to a more respectable 3.5-inch secondary battery. One final step remained to display the potential of French design.

Algérie retained the firepower of previous ships, bolstered by 3.9-inch secondaries, plus a significant range increase. Her weight of protection exceeded 2000 tons. These advances dictated some compromise, detracting from her sea-keeping and eliminating the machinery dispersal of the earlier classes. Nevertheless, her well-rounded design suited her well to France's needs and made her capable of challenging any foreign rivals, even the larger ones.

Figure 8. *Algérie* had the thickest deck protection of any wartime cruiser. (courtesy of the Boris Lemachko Collection)

TRENTO

Displacement: 10,345 tons
Range: 4160nm at 16 knots
Speed: 31 knots
Armor: 2.8-inch belt, 2-inch deck, 3.9-inch turret
Main Battery: eight 8-inch guns, 2210-lb broadside

ZARA

Displacement: 11,682 tons
Range: 5361nm at 16 knots
Speed: 31 knots
Armor: 5.9-inch belt, 2.8-inch deck, 5.9-inch turret
Main Battery: eight 8-inch guns, 2210-lb broadside

Figure 9. *Zara* and her sisters had a strong claim to being the best-protected cruisers in the war. (courtesy of History on CD-ROM)

BOLZANO

Displacement: 10,890 tons
Range: 4432nm at 16 knots
Speed: 33 knots
Armor: 2.8-inch belt, 2-inch deck, 3.9-inch turret
Main Battery: eight 8-inch guns, 2210-lb broadside

The exaggeration in Italian speed claims is now well known. At the time, high trial speeds represented a fine way to earn bonuses and publicity for domestic shipyards. The navy, fully aware of the ships' true capabilities, never banked on unrealistic expectations.

The *Trento* class, often grouped with the *Duquesne*s as tinclads, in fact had much better protection. The belt had little hope of defeating an 8-inch shell outright, but it could keep out some smaller rounds. Deck protection was not bad, and the turret armor had value against shells striking at an angle, if not against solid hits.

In contrast, the *Zara*s had protection unsurpassed by any wartime cruiser. This improvement came with little compromise, thanks to ignoring the treaty tonnage rules, but *Zara* did have to compress her propulsion into a two-shaft layout. Nevertheless, internal arrangements and subdivision remained good.

When the leadership opted for a final heavy cruiser modeled on *Trento* rather than *Zara*, poor *Bolzano* earned the nickname "Magnificent Mistake." She gained little over *Trento* while pushing the treaty limit even farther.

All seven ships shared some traits. Their Mediterranean environment made minimal demands on their sea-keeping and range, and they would not likely have prospered in a more open setting. They mounted powerful guns, especially the newer model in *Bolzano* and the *Zara*s, but these produced a fair amount of trouble. Irregular performance, aggravated by a close spacing of the muzzles, led to excessive dispersion in combat. The Italians addressed the problem by adjusting their propellants and shell weight (including 260-lb shells for a broadside of 2081 lbs). The 3.9-inch secondary guns, though getting long in the tooth, were otherwise practical. Italian fire control systems proved capable.

THE PACIFIC

The Japanese Navy's imperialistic ambitions made a showdown with the United States almost a certainty. The Pacific setting demanded long range, especially for the Americans; both fleets anticipated a final duel taking place on the western side of the ocean.

The oceanic expanse increased focus on aviation. Japanese doctrine placed special emphasis on cruiser aircraft in the scouting role; yet it was the Americans who fitted the most lavish aviation outfits, with four aircraft and two catapults in each class. Both navies fitted their heavy cruisers with heavy AA guns firing shells of 45 lbs and up, as opposed to the 30-35 lbs typical in other navies. The Americans went as far as to pull the torpedoes off their ships to help compensate for enlarged DP batteries. The experience of World War II would show that heavy cruisers had little use for torpedoes, but the Japanese had a special investment in their 24-inch weaponry, seen as an equalizer against America's superior battle line. All of Japan's heavy cruisers eventually carried "Long Lance" oxygen torpedoes,[2] though it did them almost as much harm as it did to the enemy.

FURUTAKA

Displacement: 9150 tons[3]
Range: 7000nm at 14 knots
Speed: 33.3 knots
Armor: 3-inch belt, 1.4-inch deck, 1-inch turret
Main Battery: six 8-inch guns, 1665-lb broadside

MYOKO

Displacement: 13,000 tons
Range: 8500nm at 14 knots
Speed: 33.3 knots
Armor: 4-inch belt, 1.4-inch deck, 1-inch turret
Main Battery: ten 8-inch guns, 2774-lb broadside

TAKAO

Displacement: 13,400 tons
Range: 8500nm at 14 knots
Speed: 34.25 knots

---

[2] The term "Long Lance" appears in the translation of a Japanese document captured at Kwajalein, but it's unclear if this is the origin of the nickname.
[3] Japanese cruisers underwent dramatic prewar upgrades, and the statistics given here reflect those improvements.

Armor: 5-inch belt, 1.4-inch deck, 1-inch turret
Main Battery: ten 8-inch guns, 2774-lb broadside

MOGAMI

Displacement: 12,400 tons
Range: 7500nm at 14 knots
Speed: 35 knots
Armor: 5.5-inch belt, 2.4-inch deck, 1-inch turret
Main Battery: ten 8-inch guns, 2774-lb broadside

TONE

Displacement: 11,230 tons
Range: 12,000nm at 14 knots
Speed: 35 knots
Armor: 5.7-inch belt, 2.6-inch deck, 1-inch turret
Main Battery: eight 8-inch guns, 2220-lb broadside

The *Furutaka* class and the similar *Aoba*s, already in the works at the time of the Washington Conference, would be the only Japanese heavy cruisers to comply with the treaty limits. Though described as displacing 7100 tons throughout their careers, in fact they had entered service as much as 1200 tons above that weight, and their extensive modernizations pushed them near 10,000 tons. Outwardly unremarkable, these were respectable, well-balanced ships that demonstrated admirable resistance to gunfire, if not to underwater damage.

The *Myoko*s began at almost 11,000 tons, and the *Takao*s exceeded even that figure. Some of the excess resulted from faulty design calculations, but the Japanese showed little reluctance to ignore their treaty obligations. Though no Japanese cruisers enjoyed the advantages of machinery dispersal, *Myoko* and *Takao* had a good degree of subdivision that gave them excellent, even remarkable, resistance to torpedoes. Their qualities passed on to subsequent classes.

When laid down as a fast, armored ship mounting fifteen 6.1-inch guns, *Mogami* bore the ludicrous claim of an 8500-ton displacement, though in fact her design had already grown 1000 tons beyond it, and she would not commission until she had topped 11,000. The initial low-tonnage goal accomplished nothing but to necessitate a remedial campaign for the frail, unstable vessel it had created. When all the work was finished, however, the *Mogami*s fit in well with the preceding classes.

Tone, beneficiary of all the unpleasant experience with previous classes, entered service in fully mature form. It was an odd form, with its specialization in floatplane service, and the fleet did not have any great love for it: the next planned cruiser descended from *Mogami*'s design, and no further *Tone*s were planned. Crewmen, on the other hand, enjoyed *Tone* for her relatively luxurious accommodations.

Japanese heavy cruisers carried a disappointing 8-inch gun. Its shell had a unique form that maximized performance against homogenous armor and attempted to provide a hydrodynamic profile to increase hits below the waterline. This degraded the capabilities against face-hardened armor, and the fuze setting for underwater performance often allowed direct hits to pass through a target before detonating. The guns themselves fired slowly, and thin protection left their mounts vulnerable to the smallest shells. The 5-inch secondary gun aboard most of these cruisers had the ability to throw a hefty sum of metal and probably rates second only to US 5-inch guns as a cruiser DP weapon.

Figure 10. This wartime intelligence drawings of *Nachi* blends accurate and inaccurate features. The full-height belt covers only the midships (see lower course of portholes), and the hull bulge indicates volume unprotected by armor. (courtesy of History on CD-ROM)

PENSACOLA

Displacement: 9100 tons
Range: 10,000 nm at 15 knots
Speed: 32.8 knots
Armor: 2.25-inch belt, 1.75-inch deck, 2.5-inch turret
Main Battery: ten 8-inch guns, 2600-lb broadside

NORTHAMPTON

Displacement: 9050 tons
Range: 10,000 nm at 15 knots
Speed: 32.5 knots
Armor: 2.25-inch belt, 2-inch deck, 2.5-inch turret
Main Battery: nine 8-inch guns, 2340-lb broadside

PORTLAND

Displacement: 9800 tons
Range: 10,000 nm at 15 knots
Speed: 32.7 knots
Armor: 2.25-inch belt, 2.5-inch deck, 2.5-inch turret
Main Battery: nine 8-inch guns, 2340-lb broadside

NEW ORLEANS

Displacement: 9950 tons
Range: 10,000 nm at 15 knots
Speed: 32 knots
Armor: 5-inch belt, 2.25-inch deck, 8-inch turret
Main Battery: nine 8-inch guns, 2340-lb broadside

WICHITA

Displacement: 10,565 tons
Range: 10,000 nm at 15 knots
Speed: 33.6 knots
Armor: 6-inch belt, 2.25-inch deck, 8-inch turret
Main Battery: nine 8-inch guns, 2340-lb broadside

BALTIMORE

Displacement: 13,600 tons

Range: 10,000 nm at 15 knots

Speed: 33 knots

Armor: 6-inch belt, 2.5-inch deck, 8-inch turret

Main Battery: nine 8-inch guns, 3015-lb broadside

Figure 11. *Wichita* combined powerful weaponry, thick armor, and good mobility.

When finally funded to pursue a new cruiser program, the Americans proceeding with great zeal. Zeal for observing the treaty limits resulted in a ship grossly underweight, while zeal for firepower attempted to crowd too many guns into a light hull. The *Pensacola*s couldn't take full advantage of their ten-gun batteries because their hull form provided an unsteady platform and because fire control remained more primitive than in later ships.[4] The design was recognized at once as tinclad, and though the turret faceplates could defeat most 5-inch shells (even some 6-inch shells), the barbettes were barely adequate against 8-inch splinters.

Some good features became standard in American cruiser design. The propulsion machinery performed reliably with good range, though the designed range would become decreasingly practical due to wartime weight gain. Subdivision was no more than adequate, but dispersal of the machinery enhanced survivability. Habitability was generally good.

The American 8-inch gun relied especially on crew performance to achieve adequate accuracy. However, it proved possible to exceed the designed rate of fire, matching the output of the fastest foreign models. The 5-inch/25 gun fired rapidly and flung a heavy shell, but with a very low muzzle velocity.

Even before *Pensacola* was launched, designers set about correcting perceived problems. Unfortunately, not all the problems could be perceived in the incomplete ship; for example, the extent of the underweight remained unknown, so a planned increase in armor failed to take full advantage. Still, *Northampton* represented a significant advance. Design work continued, but the rapid pace of construction allowed only limited change for the next class, the *Portland*s.

It was with the *New Orleans* class that designers managed to fully digest the experience of the preceding ships. Most of the virtues were retained, while the armor achieved thicknesses competitive with those in any foreign ship. The most significant compromise was the loss of machinery dispersal.

---

[4] American cruisers eventually mounted cutting-edge radar, but in focusing on the fundamental ship designs, this essay overlooks more "swappable" components like radar and light AA guns.

New Orleans would have made a worthy nominee for the best treaty cruiser award if not for the next American development.

Wichita had an improved 8-inch mount, wider to reduce muzzle interference and to allow mounting the guns in individual sleeves. Vertical armor was face-hardened. The secondary battery consisted of a more powerful 5-inch/38 gun. On the negative side, stability was lacking, a problem serious enough to require 200 tons of solid ballast.

The end of treaty restrictions made possible a whole new set of bold specifications, and the Americans built the *Baltimore* class with improved stability, machinery dispersal, and the other desirable family traits. In addition, the *Baltimores* went to sea with a new 8-inch shell weighing 335 lbs. *Wichita* and some *New Orleans*-class units had the equipment to handle the super-heavy shell, but while at least one ship did receive them, it appears that none got into action with them.

WORLDWIDE COMMITMENTS

Britain found itself in the awkward position of having to contemplate simultaneous naval campaigns in the Mediterranean, the North Sea, the open Atlantic, the Pacific, the Indian Ocean—virtually everywhere a cruiser could venture. The resultant need for vast numbers of hulls caused an early shutdown of treaty cruiser construction in favor of smaller types. So while the British were early entrants into the treaty cruiser race, with fifteen units (including two for Australia) completed in just four years, the timeline hardly allowed for maturation. Thus the Royal Navy is ill represented in the treaty cruiser contest.

Content with modest aviation requirements, the British usually had a single aircraft on board rather than the two or three typical in other fleets.

KENT

Displacement: 9850 tons
Range: 13,300nm at 12 knots
Speed: 31.5 knots
Armor: 1-inch belt, 3-inch deck, 1-inch turret
Main Battery: eight 8-inch guns, 2048-lb broadside

Figure 12. The "Counties" didn't have the most modern look, but their operational virtues gave the British what they needed.

YORK

Displacement: 8250 tons
Range: 10,000nm at 14 knots
Speed: 32.3 knots
Armor: 3-inch belt, 3-inch deck, 1-inch turret
Main Battery: six 8-inch guns, 1536-lb broadside

The thirteen cruisers known informally as the "County" class—more officially as the *Kent*, *London*, and *Dorsetshire* classes—differed among themselves but not to any great degree. They all possessed the virtues of superb sea-keeping along with good range and habitability, precisely the qualities needed of them. They also shared a lack of protection. Some of them eventually enjoyed the addition of belt protection for the machinery spaces, and thanks to simultaneously weight-reduction measures, this was managed without raising the displacement beyond about 10,300 tons.

Before the need for cheaper ships caused a reversion to 6-inch guns, an attempt was made to fashion a humbler 8-inch design, resulting in the near-sisters *York* and *Exeter*. These ships mounted their midships belts from the start, and they had 2-inch protection for their hoists. However, they couldn't match the sea-keeping and accommodations of the larger ships. Intended to be cheaper rather than better, they achieved that goal. The next class would have returned to an eight-gun battery and the full tonnage allowance, but finances killed the project, and with the signing of the London Treaty, attention shifted to 6-inch designs.

HIPPER

Displacement: 14,050 tons
Range: 6500nm at 17 knots
Speed: 32.5 knots
Armor: 3.1-inch belt, 2-inch deck, 6.3-inch turret
Main Battery: eight 8-inch guns, 2152-lb broadside

KIROV

Displacement: 7765 tons
Range: 3750nm at 17.8 knots
Speed: 35.94 knots
Armor: 2-inch belt, 2-inch deck, 2-inch turret
Main Battery: nine 7.1-inch guns, 1935-lb broadside

Desperate to manage emerging global threats, Britain attempted to corral Germany and the Soviet Union into the treaty system. Treaties were indeed signed, but with little appreciable effect on construction.

The Germans went ahead with a treaty cruiser that wildly exceeded the treaty limits. Despite the exorbitant size, the *Admiral Hipper* class showed little real-world advantage over foreign types. Without a doubt, *Hipper*'s finest attribute was her armament—powerful guns with accurate fire control, thickly protected to keep the weaponry in action as long as possible. Armor for the hull, though, was surprisingly poor. The three completed units had various propulsion plants, none of which performed as intended.

The closest the Soviets came to owning a treaty cruiser was the purchase of an incomplete *Hipper* from Germany, and thanks to the war, the ship never reached completion. The only other modern cruisers, the *Kirovs*, evolved entirely without regard to treaty restrictions. Of the six ships, the last four were improved with 2.8-inch belt and turret armor. But despite cramming nine guns into a small displacement, the *Kirov*s could not compare with full-sized heavy cruisers in combat or operational ability.

CONCLUSION

The question of which treaty cruiser was the best depends first on the question of which cruisers qualify as treaty designs. Clearly, ships like *Baltimore* and *Hipper* lie outside the category. Negotiators at the Washington Treaty acknowledged that completed ships might turn out with a slight excess of weight, and that their ongoing modifications might add some tons, formulating a "gentlemen's agreement" on the subject. Allowing a 5% initial overweight and a 300-ton eventual growth brings us just short of 11,000 tons.

The only French ship worth considering is *Algérie*, a ship of sufficient quality in almost all facets. The navy needing a well-rounded combatant might consider her the best treaty design.

In their fine *Zara* class, the Italians produced a design likely to excel in the confines of the Mediterranean, but the tonnage ceiling leaves only the *Trento*s and perhaps *Bolzano* in the competition. It is difficult to find criteria that qualify them as the best.

The *Furutaka*s don't equal *Algérie* in all-around quality, but they defied their age and continued as able warships twenty years after they were designed. They could be the first choice for a navy planning to make extensive use of torpedo warfare, as the Japanese fleet was. The later Japanese ships were superior, certainly, but the excellence came at the cost of excessive size.

With stability issues as her only notable vice, *Wichita* appears equal or superior to *Algérie* in almost every way. A choice between the two might come down to tastes in armor layout.

As operationally superb as the "Counties" were, only a myopic fix on sea-keeping could make them a top choice. The addition of belt armor to the *Kent*s was an important upgrade without excess tonnage, but not enough to rate them with more modern designs.

Given the individual needs of various navies, no single cruiser design qualifies as the singular superior to all others. However, the search for a ship with good all-around qualities that would perform well in general cruiser duties might well end at *Wichita*.

THAT WHICH WE CALL A CRUISER: TREATIES AND THE VOCABULARY OF SHIP DEVELOPMENT

An event as momentous as the Second World War leaves our memories with a set of defining images that help us understand the subsequent direction the world has taken. In some cases, these new definitions distance us from pre-war thought, and hindsight obscures rather than explains the realities of that time. Such is the case of the word *cruiser*, which has undergone multiple shifts in meaning.

At the end of the 19th Century, international naval convention used the cruiser label for warships just below battleships in size and prestige. They were multi-role ships whose duties included scouting for the fleet, countering torpedo craft, and executing commerce warfare. The label accumulated a set of adjectives based on the ship's size (3rd class cruiser), primary role (scout cruiser), or armor scheme (protected cruiser).

Japanese Cruiser "Akashi"　石　明　艦洋巡國帝

Figure 13. Improvements in metallurgy made it possible to fit belt armor on high-speed ships, making protected cruisers like *Akashi* obsolete. (courtesy of the Boris Lemachko Collection)

**Figure 14. Armored cruisers could have the size of a battleship, as
North Carolina demonstrates.**

One type of cruiser broke out of the pack. The armored cruiser (so
named because it had belt armor) grew to great size, even exceeding that of
some battleships, and it took on a capital-ship role like that of a 2nd-class
battleship. So elevated did its status become that it earned a new label, the
battlecruiser. The smaller cruisers—and they were indeed smaller, as shown
in 1913 when the British launched the 28,000-ton *Tiger* and the 3750-ton
Arethusa—settled into a common "light cruiser" category. After the Great
War, when the time came for the Royal Navy discuss its next generation of
weaponry, the Principle Question appeared under the heading "Armaments
for Future Capital Ships and Light Cruisers"—the two general categories for
large surface combatants. But generalities were about to be strangled by the
strictures of international diplomacy.

The Five-Power Treaty signed in Washington in 1922 imposed a specific
vocabulary on naval construction.

> *No vessel of war exceeding 10,000 tons (10,160 metric tons)*
> *standard displacement, other than a capital ship or aircraft-*

carrier, shall be acquired by, or constructed by, for, or within the jurisdiction of, any of the Contracting Powers....

No vessel of war of any of the Contracting Powers hereafter laid down, other than a capital ship, shall carry a gun with a calibre in excess of 8 inches (203 millimetres).

These restrictions from Articles XI and XII of the Washington Treaty seem straightforward at first glance, but they warrant a closer look. A primary aim of the treaty was to limit the numbers of capital ships serving in the major fleets. Pursuant to this, it had to formalize the term *capital ship*, previously an informal reference to battleships and the largest cruisers. Consequently, even though the treaty never used the word *cruiser*, it effectively split the cruiser family into two branches: the massive battlecruisers, capital ships whose construction would be strictly regulated; and the light cruisers which could be built in unlimited numbers within the 10,000-ton and 8-inch limits—the "treaty cruisers."

Common knowledge will, with great sincerity, define *treaty cruiser* as nearly synonymous with *heavy cruiser*, and so it eventually came to be. However, as understood at the Washington Conference, the treaty cruiser was a light cruiser. The heavy cruiser did not even exist at the time, at least not as a specific type of warship. Light cruisers were "light" to distinguish them, not from "heavy cruisers" *per se*, but from the armored cruisers; the light cruiser had a thin armor belt and thus vaguely resembled a miniaturized armored cruiser. Not until the 1930's would the heavy cruiser become a standard classification, but hindsight has stretched this usage back through time.

Why does this matter? The labels themselves have no great importance, and in any case, there's no reconciling the minutiae of each navy's peculiar rating system. However, for folks looking back at the ships of this period, a misunderstanding of the warship typology will obscure the process of design development. The widespread assumption that heavy cruisers descended from the armored cruiser makes the evolution of the treaty cruiser seem like a twisted joke, yet the decisions begin to make sense when we identify the light cruiser as the starting point.

In World War I, the typical light cruiser was a 4000-ton ship with belt armor at 3-inch maximum thickness and a handful of 6-inch guns in single, shielded mounts scattered around the deck; but things began to change. In 1918, the British laid down their two "E" class ships displacing over 7000 tons, one of them equipped with a twin turret. The Americans started their *Omaha* class the same year, similar to the "E" class in size, with all 6-inch guns in enclosed mounts including two twin turrets. But an even more significant design had entered service by then; Britain's *Cavendish* class displaced 9000 tons, and its single, shielded mounts carried 7.5-inch guns.

None of these designs had pretensions to capital-ship status; all three were officially rated as light cruisers. The resemblance between *Cavendish* and the ensuing "treaty cruiser" is unmistakable, and it's clear that the treaty limits were laid out in part to accommodate that class. However, there were other factors. As the Washington Conference was getting underway, the Japanese formulated their requirements for the *Furutaka* class: a 7100-ton design with 3-inch belt armor and 7.9-inch guns all in turrets. It's not surprising that the *Furutaka*s came to be known as treaty cruisers, and they fit in fairly well with the post-treaty designs, though history shows their conception preceded the treaty.

With its generous limits, the treaty allowed for light cruisers enlarged to accommodate the latest international fashions, and upon signing the treaty, some countries decided there was no worthwhile cruiser that did not fill the 10,000-ton, 8-inch allowance. The Americans, having just commissioned their *Omaha*s, suddenly decided they could not bear with anything short of the full treaty figures; they built nothing but 10,000-ton, 8-inch cruisers until forced by later treaties to try something else. Likewise, the Japanese built to the treaty limit (and beyond it) with such zeal that they pulled the 7.9-inch guns out of their cruiser force for replacement by 8-inch guns—a gain of 3 millimeters. The British, with their traditional status as the premier cruiser power, gamely tried to keep up, but their realization that such a course would be ruinously expensive helped set the stage for the next round of naval diplomacy, and the next set of cruiser definitions.

Regarding arms limitation, the multi-purpose nature of cruisers posed a problem fundamentally different from that posed by battleships. Battleships had one over-riding mission, to control the seas, a role best handled by concentration of all available units into a single fleet; the fleet with the most battleships was the strongest. Cruisers, though, had a habit of skulking independently throughout the seas, causing mischief in remote sectors even if the enemy boasted an overwhelming superiority in cruiser numbers. The British Empire's far-flung lines of communication presented innumerable remote sectors for enemy mischief-makers, and to secure those sectors, the Royal Navy needed cheap designs it could build in large numbers. In contrast, the Americans needed individually large ships that could carry a war across the vast Pacific from California to the Philippines.

The Washington Treaty signatories, buoyed by their success in limiting capital ships, hoped to apply similar restrictions on "auxiliary" ships like cruisers. When diplomats gathered in Geneva in 1927, the quantity-quality cruiser dichotomy initiated a tortuous process of wordsmithery. Some negotiators tried to distinguish large "offensive" cruisers from small "defensive" cruisers. More colorful language depicted "fighting" cruisers and "peace police" cruisers. Ultimately, the vocabulary calmed down to a bland "heavy" versus "light" distinction. In using these terms, negotiators subtly altered what a "light" cruiser was—no longer a "lightly armored cruiser," it now distinguished designs utilizing less than the full 10,000-ton allotment.

This definition lasted but a moment. Because a smaller cruiser could not mount guns larger than 6-inch and remain a well-balanced design, the light/heavy distinction became synonymous with the 6-inch/8-inch gun distinction.

Simple as this was, it hardly made the treaty process an easy one because there was no international consensus comparing the combat value of 6-inch and 8-inch cruisers. Negotiators labored for a "yardstick" by which they could declare that X tons of 6-inch cruisers equaled Y tons of 8-inch cruisers. The Geneva proceedings broke down on the cruiser issue, but diplomatic persistence paid off in the 1930 London Treaty. Signatories received tonnage quotas for cruisers of two specific types defined in Article 15.

The cruiser category is divided into two sub-categories, as follows:

(a) Cruisers carrying a gun above 6.1 inch (155 mm) calibre;

(b) Cruisers carrying a gun not above 6.1 inch (155 mm) calibre.

Just as the Washington Treaty never used the word "cruiser," the First London Treaty never mentions "light" or "heavy" cruisers. However, the convention equating "light" with 6-inch guns and "heavy" with 8-inch guns was firmly established. (Note that, in raising the light cruiser gun caliber to 6.1 inches, the treaty acknowledged France's recent completion of small cruisers with 155mm guns. The Japanese saw this 3mm increase and, true to form, abandoned the 6-inch caliber that had been a fleet standard since the preceding century, building their next light cruisers with 155mm guns.)

Figure 15. *Louisville* **began as a light cruiser but became a heavy cruiser—not by reconstruction but by treaty ink.**

The new terminology had several results. First, it obscured the fact that heavy cruisers were actually light cruisers that happened to be armed with 8-inch guns. Since many armored cruisers—the ships that light cruisers had formally been distinguished from—also carried 8-inch guns, it became easy to mistake heavy cruisers for a modern form of armored cruiser. However, armored cruisers routinely carried armor in thicknesses of 6-inch or more, while some heavy cruisers had no plating thicker than 30mm; readers do well to remember that the two types had little to do with each other.

The Americans unwittingly increased the confusion. They formalized the new distinction by dividing their cruiser fleet into the heavy and light categories even though this necessitated shifting some light cruisers into the new heavy cruiser category. For example, USS *Louisville*'s hull number went from CL-28 to CA-28. The fact that CL signified a light cruiser is obvious enough, but for some reason the Americans opted against a CH prefix for their heavy cruisers, instead borrowing the old CA prefix used formerly for armored cruisers. *Montana*, the last American armored cruiser, had 9-inch turret armor; *Louisville* had 2.5-inch turret armor.

The British were not complicit in this obfuscation. They rated all their cruisers, from the treaty period all the way through World War II, simply as "cruisers." If pressed to make the distinction, they would likely respond with "8-inch cruiser" and "6-inch cruiser."

But the damage had been done, and struggles with classification extended beyond the English language. The Italians had no specific cruiser rating until 1929, after the cruiser question had wrecked the Geneva Conference. They then, like the Americans, had to do some light-to-heavy juggling; the first 10,000-ton ships appeared under the heading of "Type A light cruisers" (*incrociatori leggeri categoria "A"*), but other adjectives quickly stepped in: "heavy" (*pesanti*) and "greater" (*maggiori*) and even "armored" for some later ships (*corazzati*, not a term the Italians had ever applied officially to their old armored cruisers). The 6-inch classes rated as "large scouts" (*grandi esploratori*), then "Type B light cruisers" until the "Type B" qualifier became unnecessary. During the war, the small "Capitani Romani" cruisers started with the label "ocean-going scouts" (*esploratori oceanici*) and later slipped into the light cruiser category, and the incomplete *Etna* and *Vesuvio* were *incrociatori antiaerei*.

The Soviets had an uncomplicated approach—all their cruisers were "cruisers." Unofficial correspondence could sometimes affix the "light" adjective (*legkie kreisera*) even to ships those mounting 7.1-inch guns, while "heavy cruisers" (*tiazhelie kreisera*) included more powerfully-armed units, especially battlecruisers—a common point of misunderstanding in Soviet-era accounts.

Figure 16. The 180mm battery put *Kirov* in the treaty category of heavy cruisers, but the Soviets used the term differently. (courtesy of the Boris Lemachko Collection)

There was at first a similar simplicity with the Germans, as they constructed 5.9-inch "cruisers." Then came the large commerce raiders dubbed "armored ships" (*panzerschiffe*) in reference to Versailles Treaty nomenclature. During the war, the rating changed to "heavy cruisers" (*Schwere Kreuzer*), a more appropriate title that had also been applied to Germany's 8-inch ships. The final, incomplete 5.9-inch projects were designated "light cruisers" (*Leichte Kreuzer*) and "reconnaissance cruisers" (*Spähkreuzer*).

Perhaps the most straightforward system belonged to the French, who started their interwar cruiser construction with a simple *croiseur* label. When

Emil Bertin was built as a cruiser-minelayer, she got a *croiseur mouilleur de mines* rating. The purpose-built training cruiser *Jeanne d'Arc* was a *croiseur-école*. And when *Suffren* introduced higher levels of protection, she became a *croiseur protégé*, a title the French had not adopted for their 19th Century protected cruisers. In most cases, a displacement figure was appended, making *Algérie* a "protected cruiser of 10,000 long tons" (*croiseur protégé de 10.000 tonnes W*, the "W" referencing Washington). The tonnage made the only distinction from the light cruiser *La Galissonnière* (*croiseur protégé de 7.720 tonnes W*).

The Washington Treaty found the Japanese busy with ships provisionally divided into small-type (*ko-gata junyōkan*), medium-type (*chū-gata junyōkan*), and large-type (*ō-gata junyōkan*) cruiser categories, the last being those mounting 7.9-inch guns and treated to official ratings as first-class cruisers (*ittō junyōkan*) and "A class" cruisers (*kō-kyū junyōkan*)—in time, unofficially heavy cruisers (*jū-junyōkan*). The smaller ships with their 5.5-inch guns rated as second-class cruisers (*nitō junyōkan*) or unofficially as light cruisers (*kei-junyōkan*). When the *Mogamis* came along after the London Treaty, they presented a quandary, being large but carrying 6.1-inch guns. The Japanese consequently changed their classification system, basing it on gun caliber (exactly correlating to the London criterion) instead of displacement. *Mogami* ended up as a second-class or "B class" cruiser (*otsu-kyū junyōkan*), informally a light cruiser, until rearmed with 8-inch guns. A single "C class" cruiser (*hei-kyū junyōkan*, the submarine squadron flagship *Oyoda*) reached completion, but a 5800-ton AA cruiser (*bōkū junyōkan*) failed to progress into a construction order. A few projects had unusual ratings. *Katori* was classed as a training cruiser (*renshū junyōkan*, actually not considered a subset of the cruiser type). The small cruisers captured from China were rated first as escort vessels (*kaibōkan*, literally "coast defense ships"), with one unit surviving long enough to become a second-class cruiser.

The "A class" and "B class" categories look suspiciously like definitions "a" and "b" from Article 15 of the London Treaty (cited above), but it's mere coincidence. Divisions within the treaty clauses routinely received alphabetic headings. The Japanese established their A-B distinction on the basis of displacement decades before the treaties, the dividing line being 7000 tons.

(Japan's Tsushima-era armored cruisers displaced between 7600 and 9700 tons.) Again by coincidence, the "light" (tonnage) cruisers in Geneva discussions typically displaced about 7000 tons.

The systems of classification give an interesting glimpse into administration in the various navies. Readers must keep in mind, however, that nomenclature makes a dubious guide to design evolution, especially amid the differing practices in the world's navies.

APPENDIX: FURTHER READING

Warship enthusiasts quickly learn that cruisers don't enjoy the fascination lavished upon battleships, as a glance at *Books in Print* will prove. However, the interested reader has several worthy titles to pursue. *Cruisers of World War Two* by M. J. Whitley gives the best encyclopedic treatment. While it includes no armor schematics, the text goes significantly beyond statistics in depicting the design realities. (Whitley covers the German pocket battleships in his *Battleships of World War Two*, and this represents his final treatment of that confusing class.) A more general reference, *Conway's All the World's Fighting Ships* (specifically the two volumes covering 1906-1921 and 1922-1946), provides plenty of dates and statistics on cruisers and most other vessels of war. My own book *Fleets of World War II* complements *Conway's* well, delving past the statistics to discuss the real-world strengths and weaknesses of the designs. Further details are available in several country-specific books.

The best American reference is the *U.S. Cruisers* volume of Norman Friedman's design history series. Unfortunately, the book neglects armor schematics almost entirely.

British Cruisers of World War Two by Raven and Roberts stands out as the premier guide to British and Commonwealth ships, though again, it lacks armor schematics.

Arguably the finest warship reference book ever published, *Japanese Cruisers of the Pacific War* by Lacroix and Wells covers its topic from almost every imaginable angle. Armor and other features are shown in detail for every modern class.

Readers will find no English-language volume on French cruisers, but the *Warship* annual often has material of interest: for example, *Warship 2001-2002*, *Warship 2005*, and *Warship 2006* cover the treaty cruisers.

The situation is worse regarding Italian cruisers. Fortunately, you don't need to know Italian to understand the armor schematics in *Gli incrociatori italiani 1861-1970* by Giorgerini and Nani, and it covers all the World War II units.

As with the German navy in general, German cruisers enjoy more attention than they deserve. Koop and Schmolke have provided *Pocket Battleships of the Deutschland Class*, *Heavy Cruisers of the Admiral Hipper Class*, and *German Light Cruisers of World War II*, all of which contain armor schematics. *German Cruisers of World War Two* rates as one of Whitley's finest works. Whitley is consistent in that he put the pocket battleships in his other book, *German Capital Ships of World War Two*.

For Soviet cruisers, it may be that the best source is *Raising the Red Banner*, a general reference on Soviet warships from 1922 to 1945, written by me and Vladimir Yakubov. We also have an article on the *Kirov* class scheduled for *Warship 2009*.

There are few options for readers interested in the smaller navies. The Dutch have at least some coverage: Quispel's *The Job and the Tools* has a good drawing for *Tromp*, and Van Oosten wrote a volume on *De Ruyter* for the Warship Profile series. Magazines occasionally feature relevant material, though the best sources are often foreign-language. For example, the September 2007 issue of *Storia Militare* has an article by Maurizio Brescia on Argentina's *25 de Mayo* class, including a single armor sketch. Hunting for sources like these involves considerable effort for little return, but true cruiser fans will find the investment worthwhile.

Figure 17. *Prinz Eugen* postwar in American hands, useful only for tests and for lugging a bomber across the Atlantic.

A LETTER FROM THE PUBLISHER

Dear Reader,

I cut my teeth as a naval history aficionado on Samuel Eliot Morison's *History of United States Naval Operations in World War II,* and at a very early age I had memorized the order of battle of the U.S. Navy at just about every battle in the Pacific War.

I knew the names and numbers of the four Allied CAs that were lost at Savo Island and I knew that the *Brooklyn* class CLs had 15 (!) 6-inch guns. I knew that the Japanese CAs were fearsome and equipped with the dreaded Long Lance torpedo I knew that the German CA *Prinz Eugen* accompanied *Bismarck* on her ill-fated final sortie. I loved playing my friend Steve Marzen in a Simulations Publications Incorporated (SPI) game called "CA", which I rarely lost—unusually for me, as I have never been much of a games tactician.

I never realized, though, what a profound impact the U.S. Navy's cruiser nomenclature had on me. Because the U.S.used the distinct acronyms "CA" and "CL" to describe what are commonly called "heavy" and "light" cruisers, I unthinkingly internalized an assumption that the two types were quite distinct in design and purpose. It is only now—forty years later!—that I learn that the differences in nomenclature were in many ways rather arbitrary and that for many purposes the simple "cruiser" is as good a term as any.

This is unquestionably the most informative and thought-provoking book I have ever read about cruisers. I am proud to bring it to you.

—*Fred Zimmerman, Nimble Books LLC,*
Ann Arbor, Michigan, USA, 2008

www.ingramcontent.com/pod-product-compliance
Lightning Source LLC
Chambersburg PA
CBHW042000100426
42813CB00019B/2943